IGNITION:

THE OVERVIEW OF COMBUSTION ENGINE

BY

ROBERT NYM

TABLE OF CONTENT

Introduction to Combustion Engines

Combustion engines, also known as internal combustion engines, are the heartbeat of modern transportation and numerous industrial applications. These remarkable machines have revolutionized the way we travel, work, and live. At their core, combustion engines are ingenious devices that convert chemical energy stored in fuel into mechanical energy, propelling vehicles and powering various machinery.

Along the line, we will delve into the fundamental concepts of combustion engines, exploring their origins, significance, and the underlying principles that make them indispensable in today's world.

1.1 Definition and Basic Principles:

At its essence, a combustion engine is a mechanical device that burns fuel within a confined space to generate power. This power, harnessed through a series of intricate processes, drives the wheels of automobiles, propels ships, and generates electricity in power plants.

1.2 Historical Evolution:

The journey of combustion engines dates back to the 19th century, with inventors like Nikolaus Otto and Rudolf Diesel pioneering the internal combustion process. We will explore the pivotal moments and inventors that shaped the development of these engines, leading to the sophisticated designs we see today.

1.3 Importance and Applications:

Combustion engines are the lifeblood of various sectors, including automotive, aviation, marine, and industrial manufacturing. The power cars, trucks, airplanes, ships, and even electricity generators. Understanding their significance is crucial to appreciating their role in shaping modern society.

Throughout this chapter, we will unravel the intricate workings of combustion engines, setting the stage for a comprehensive exploration of their components, processes, and impact on the world around us. Join us on this journey as we uncover the mysteries behind these powerful machines.

Chapter 1

Historical background of combustion engine

The history of the combustion engine is a fascinating journey that spans centuries, marked by continuous innovation and ingenuity. Here's a concise overview of its evolution:

Early Concepts (17th-18th Century)

1678: The concept of internal combustion is first documented by Christian Huygens, a Dutch scientist.

18th Century: Various inventors, including Jean Lenoir and Robert Street, experiment with atmospheric engines and demonstrate the potential of burning gases for power.

Early Practical Engines (19th Century)

1807: François Isaac de Rivaz, a Swiss inventor, develops the first hydrogen-powered vehicle, considered the first internal combustion engine automobile.

1824: Samuel Brown patents an internal combustion engine that runs on hydrogen gas.

1859: Belgian engineer Etienne Lenoir builds the first commercially successful internal combustion engine, running on illuminating gas and later coal gas.

1860s: Nikolaus Otto, a German engineer, develops the four-stroke engine, also known as the Otto cycle, which becomes the foundation for most internal combustion engines.

Rise of the Automobile (Late 19th-Early 20th Century)

1885-1886: Karl Benz and Gottlieb Daimler, working independently in Germany, develop practical petrol-powered vehicles, marking the birth of the modern automobile.

Late 19th Century: Rudolf Diesel invents the diesel engine, which operates on a high compression ratio, making it more fuel-efficient than petrol engines.

Early 20th Century: Mass production techniques, advancements in fuels, and improvements in engine design lead to the widespread adoption of internal combustion engines in automobiles and industrial applications.

World Wars and Industrialization (Early-Mid 20th Century)

World War I and II: Internal combustion engines power military vehicles, aircraft, and ships, playing a significant role in both world wars and driving further technological advancements.

Mid-20th Century: Continuous refinement and optimization of internal combustion engines lead to increased power, efficiency, and reliability. Racing competitions drive innovations, enhancing engine performance and design.

Environmental Challenges and Innovations (Late 20th Century-Present) 1970s: Environmental concerns lead to the development of emission control technologies, such as catalytic converters, to reduce harmful exhaust emissions.

Late 20th Century: Advancements in electronics and computer control systems enhance engine efficiency, enabling the introduction of fuel injection and electronic ignition systems.

21st Century: Continued emphasis on reducing emissions results in the development of hybrid vehicles, electric cars, and advancements in alternative fuels. Internal combustion engines coexist with electric powertrains, contributing to diverse transportation options.

The history of the combustion engine reflects humanity's relentless pursuit of progress, from early experimental designs to the sophisticated, high-performance engines of today. As the world moves toward a more sustainable future, the combustion engine continues to evolve, integrating cutting-edge technologies to meet environmental challenges while maintaining its essential role in various industries.

Chapter 2

Internal Combustion Engine Basics

Internal combustion engines, the heart of most vehicles and machinery, operate on the principles of harnessing controlled explosions to generate mechanical power. In this chapter, we will dissect the core components and workings of internal combustion engines, shedding light on their intricate mechanisms.

2.1 Components of Internal Combustion Engines

At its core, an internal combustion engine comprises several vital components, including pistons, cylinders, crankshafts, camshafts, valves, and a fuel delivery system. Understanding the functions of each part is essential to grasp the engine's operation.

2.2 Working Principles

Internal combustion engines operate through a series of precise steps. We will explore the four-stroke and two-stroke cycles, elucidating the intake, compression, power, and exhaust phases. These cycles are the building blocks of engine operation, governing how fuel is burned and converted into mechanical energy.

2.3 Types of Internal Combustion Engines

There are two primary types of internal combustion engines: petrol engines and diesel engines. Petrol engines rely on spark plugs to ignite the air-fuel mixture, while diesel engines use high compression to achieve ignition. We will delve into the distinctions between these engines, highlighting their applications and advantages.

2.4 Thermodynamics of Internal Combustion

A deeper understanding of thermodynamics, including concepts like entropy, pressure-volume diagrams, and heat transfer, is crucial in comprehending the efficiency and performance of internal combustion engines. We will explore these concepts, linking them to engine behavior and efficiency.

2.5 Engine Balancing and Vibration Control

Internal combustion engines inherently produce vibrations due to reciprocating and rotating components. Engine balancing techniques, such as counterweights and balance shafts, are employed to minimize these vibrations. We will discuss the importance of engine balance and methods used to control vibrations.

By the end of this chapter, readers will have a solid grasp of the fundamental components, processes, and principles governing internal combustion engines. This knowledge forms the foundation for exploring advanced topics, including engine cycles, combustion processes, and efficiency optimization techniques in subsequent chapters. Join us as we unravel the inner workings of these remarkable machines.

Chapter 3

Engine Cycles

Internal combustion engines operate on specific thermodynamic cycles that dictate their efficiency and performance. In this chapter, we will explore the fundamental engine cycles: the Otto cycle for petrol engines and the Diesel cycle for diesel engines.

3.1 Understanding the Otto Cycle

The Otto cycle, also known as the four-stroke cycle, is the foundation for most petrol engines. It comprises four stages: intake, compression, power, and exhaust. We will break down each stage, explaining the processes involved and the changes in pressure and volume throughout the cycle. Real-life applications and examples of engines utilizing the Otto cycle will be discussed.

3.2 Delving into the Diesel Cycle

The Diesel cycle, prevalent in diesel engines, varies from the Otto cycle primarily in the method of ignition. Diesel engines rely on high compression to ignite the air-fuel mixture, achieving a different set of thermodynamic processes. We will explore the intricacies of the Diesel cycle, emphasizing its advantages and applications in various vehicles and industries.

3.3 P-V Diagrams and Analysis

Pressure-volume (P-V) diagrams are invaluable tools for visualizing engine cycles. We will delve into the construction and interpretation of P-V diagrams, showcasing how they represent the thermodynamic processes within the engine. Through detailed analysis, readers will gain insights into the efficiency and performance of different engine cycles.

3.4 Efficiency and Factors Affecting It

The efficiency of an engine cycle is a critical parameter that determines how effectively it converts fuel energy into mechanical work. We will discuss the factors influencing engine efficiency, such as compression ratio, air-fuel ratio, and heat losses. Understanding these factors is essential for engineers and enthusiasts seeking to optimize engine performance.

By the end of this chapter, readers will have a comprehensive understanding of the Otto and Diesel cycles, enabling them to analyze and compare different engines based on their thermodynamic principles. The knowledge gained here will serve as a solid foundation for exploring advanced topics related to engine design, performance tuning, and emissions control in subsequent chapters. Join us as we unravel the complexities of engine cycles, illuminating the path to mastering internal combustion engine technology.

Chapter 4

Engine Components

Internal combustion engines are intricate assemblies of various components working in harmony to generate power. In this chapter, we will dissect the essential engine components, understanding their functions and significance in the overall engine operation.

4.1 Pistons and Cylinders

Pistons are key components responsible for converting the energy from combustion into mechanical motion. We will explore piston design, materials, and their movement within cylinders. The interaction between pistons and cylinders is fundamental to engine performance.

4.2 Crankshaft and Connecting Rods

Crankshafts transform the linear motion of pistons into rotary motion, driving the engine's output. Connecting rods play a crucial role in this process, linking pistons to the crankshaft. We will delve into the design principles of crankshafts and connecting rods, emphasizing their durability and balance requirements.

4.3 Valves and Camshafts

Valves control the intake of air and fuel and the expulsion of exhaust gases from the combustion chamber. Camshafts, linked to the crankshaft, regulate the opening and closing of valves. Understanding valve types, timing, and profiles, as well as camshaft design, is vital for optimizing engine performance.

4.4 Fuel Delivery Systems

Carburetors, fuel injectors, and electronic fuel injection systems ensure the precise delivery of fuel into the combustion chamber. We will explore these systems, highlighting their mechanisms, advantages, and advancements in fuel injection technology. Achieving the correct air-fuel mixture is essential for engine efficiency and emissions control.

4.5 Ignition Systems

Spark plugs and other ignition components are responsible for igniting the air-fuel mixture within the combustion chamber. We will discuss the types of spark plugs, ignition timing, and electronic ignition systems. Proper ignition ensures efficient combustion and optimal engine operation.

4.6 Cooling and Lubrication Systems

Engines require cooling to dissipate excess heat generated during combustion. Air and liquid cooling systems, along with radiators and water pumps, maintain the engine's temperature within the optimal range. Lubrication systems, including oil pumps and filters, ensure the smooth movement of engine components, reducing friction and wear.

Understanding these engine components and their interactions is crucial for diagnosing issues, performing maintenance, and optimizing engine performance. By the end of this chapter, readers will have a comprehensive knowledge of the building blocks of internal combustion engines, setting the stage for further exploration into advanced engine technologies and applications. Join us as we unravel the complexities of engine components, illuminating the path to mastering the inner workings of these remarkable machines.

Chapter 5

Fuel Injection and Ignition Systems

Fuel injection and ignition systems are integral to the efficient operation of internal combustion engines. In this chapter, we will delve into the intricacies of these systems, understanding their mechanisms, advancements, and significance in modern engine technology.

5.1 Carburetors vs. Fuel Injectors

Carburetors, once common in older engines, have been largely replaced by electronic fuel injection (EFI) systems. We will explore the differences between carbureted and fuel-injected engines, emphasizing the advantages of fuel injectors in terms of precise fuel delivery, improved efficiency, and reduced emissions.

5.2 Electronic Fuel Injection (EFI) Systems

EFI systems utilize sensors and electronic control units (ECUs) to precisely measure and deliver the optimal air-fuel mixture to the combustion chamber. We will dissect the components of

EFI systems, including sensors (such as oxygen sensors and mass airflow sensors), fuel injectors, and the ECU. Understanding these elements is crucial for achieving optimal engine performance and fuel efficiency.

5.3 Direct Fuel Injection (DFI) Systems

Direct fuel injection systems inject fuel directly into the combustion chamber, bypassing the intake manifold. This technology enhances fuel atomization and combustion efficiency, resulting in improved power output and reduced emissions. We will explore the design and benefits of DFI systems, showcasing their applications in high-performance and fuel-efficient engines.

5.4 Ignition Systems and Spark Plugs

Ignition systems, responsible for igniting the air-fuel mixture, have evolved from traditional mechanical systems to advanced electronic ignition systems. We will discuss the components of ignition systems, including ignition coils, distributors, and spark plugs. Additionally, we will explore the importance of spark plug design and materials in ensuring reliable ignition and optimal engine performance.

5.5 Timing and Control

Precise timing of fuel injection and ignition events is critical for engine efficiency and power output. We will delve into the concepts of ignition timing and fuel injection timing, exploring how engine control modules (ECMs) and sensors work together to optimize these parameters based on various operating conditions.

By the end of this chapter, readers will have a deep understanding of fuel injection and ignition systems, enabling them to grasp the technological advancements driving modern internal combustion engines. This knowledge is essential for engineers, mechanics, and enthusiasts alike, providing insights into the intricacies of engine control and performance optimization. Join us as we unravel the complexities of these critical systems, illuminating the path to mastering the advanced technologies shaping the future of internal combustion engines.

Chapter 6

Combustion Process

The combustion process lies at the heart of internal combustion engines, where the controlled explosion of fuel and air generates the mechanical energy that drives vehicles and machinery. In this chapter, we will explore the intricacies of the combustion process, understanding the factors influencing combustion efficiency and its impact on engine performance.

6.1 Air-Fuel Mixture

Combustion begins with the precise mixing of air and fuel within the combustion chamber. We will discuss the significance of the air-fuel ratio, exploring the concept of stoichiometry and the ideal ratio for complete combustion. Achieving the correct mixture is crucial for maximizing energy release during combustion.

6.2 Ignition and Flame Propagation

Once the air-fuel mixture is compressed within the combustion chamber, it is ignited by the spark plug (in petrol engines) or high compression (in diesel engines). We will delve into the ignition process, understanding the formation and propagation of the flame front. Factors influencing ignition timing and flame stability will be explored in detail.

6.3 Combustion Phases

Combustion occurs in distinct phases, including ignition delay, rapid combustion, and burnout. Each phase plays a vital role in the overall efficiency and power output of the engine. We will analyze these phases, examining the factors affecting the duration and intensity of each phase. Insights into combustion chamber design and piston geometry will be provided.

6.4 Combustion Efficiency and Emissions

The efficiency of the combustion process directly impacts engine performance and fuel economy. We will explore techniques for improving combustion efficiency, such as optimizing air intake, enhancing turbulence, and employing advanced fuel injection strategies. Additionally, we will discuss the formation of emissions during combustion, highlighting the importance of emission control technologies.

6.5 Detonation and Knocking

Detonation and engine knocking are undesirable phenomena that can damage the engine and reduce efficiency. We will examine the causes of detonation and knocking, exploring methods to prevent these issues, including the use of higher-octane fuels, engine design modifications, and electronic engine control strategies.

By the end of this chapter, readers will have a comprehensive understanding of the combustion process, from the initial mixture preparation to the final expulsion of exhaust gases. This knowledge is essential for engineers and enthusiasts seeking to optimize engine performance, improve fuel efficiency, and reduce emissions. Join us as we unravel the complexities of combustion, illuminating the path to mastering the science behind the power within internal combustion engines.

Chapter 7

Engine Cooling and Lubrication Systems

Effective engine cooling and lubrication are paramount for maintaining internal combustion engines' optimal performance and longevity. In this chapter, we will explore the intricate systems designed to regulate engine temperature and reduce friction, ensuring smooth operation and durability.

7.1 Engine Cooling Systems

Engines generate significant heat during combustion, making cooling systems crucial for dissipating this excess heat. We will discuss the principles behind air cooling and liquid cooling systems, exploring components such as radiators, water pumps, fans, and thermostats. The importance of maintaining the right operating temperature and methods for preventing overheating will be emphasized.

7.2 Liquid Cooling Systems

Liquid cooling systems, common in most modern vehicles, utilize coolant (usually a mixture of water and antifreeze) to absorb heat from the engine. We will delve into the components of liquid cooling systems, including the radiator, water pump, hoses, and the thermostat. Proper circulation and heat exchange within the system are essential for efficient engine cooling.

7.3 Air Cooling Systems

Air-cooled engines, found in certain motorcycles and small aircraft, rely on natural or forced airflow to dissipate heat. We will explore the design principles of air-cooled engines, discussing finned cylinders and cooling fans. Understanding the limitations and advantages of air cooling systems is vital for selecting appropriate applications.

7.4 Engine Lubrication Systems

Lubrication systems reduce friction between moving engine components, preventing wear and ensuring smooth operation. We will discuss the functions of engine oil, oil pumps, filters, and the oil pan. Proper lubrication is essential for extending engine life, and we will explore the importance of viscosity, additives, and oil change intervals.

7.5 Friction Reduction and Wear Prevention

Reducing friction is critical for engine efficiency. We will explore methods employed to minimize friction, including engine coatings, high-quality lubricants, and advanced materials. Additionally, we will discuss the role of oil additives and filtration in preventing wear and maintaining engine cleanliness.

By the end of this chapter, readers will have a comprehensive understanding of engine cooling and lubrication systems, enabling them to appreciate the engineering behind these vital components. Proper maintenance and understanding of these systems are essential for ensuring the longevity and reliability of internal combustion engines, making this knowledge invaluable for engineers, mechanics, and enthusiasts alike. Join us as we unravel the complexities of engine cooling and lubrication, illuminating the path to mastering the art and science of engine maintenance.

Chapter 8

Engine Performance and Efficiency

Engine performance and efficiency are paramount in the design and operation of internal combustion engines. In this chapter, we will explore the measurements, factors, and techniques that influence engine performance, power output, and fuel efficiency.

8.1 Power and Torque

Understanding power and torque is essential for evaluating an engine's performance. We will explore the definitions of power and torque, their relationship, and how they affect a vehicle's acceleration and overall drivability. Methods for measuring and calculating power and torque will be discussed.

8.2 Specific Fuel Consumption (SFC)

Specific fuel consumption quantifies an engine's efficiency by measuring the amount of fuel consumed per unit of power produced. We will delve into the concept of SFC, exploring its significance in differentiating between efficient and inefficient engines. Factors influencing SFC, such as combustion efficiency and mechanical losses, will be analyzed.

8.3 Factors Affecting Engine Efficiency

Several factors impact engine efficiency, including compression ratio, air-fuel ratio, and mechanical losses. We will discuss the ideal conditions for achieving maximum efficiency and methods for optimizing these factors. Engine design modifications and advancements, such as variable valve timing and turbocharging, will be explored in the context of efficiency improvement.

8.4 Emission Control Systems

Efficient engines also need to address environmental concerns. We will discuss emission control systems, such as catalytic converters and exhaust gas recirculation (EGR), which play a vital role in reducing harmful emissions. The balance between engine performance and emission control will be explored, highlighting the importance of meeting stringent environmental standards.

8.5 Performance Tuning and Modifications

Enthusiasts often seek ways to enhance engine performance. We will explore performance tuning techniques, including modifications to intake and exhaust systems, engine mapping, and forced induction. While these modifications can increase power output, they also require a delicate balance to maintain reliability and efficiency.

By the end of this chapter, readers will have a comprehensive understanding of engine performance metrics, efficiency measurements, and the intricate balance between power output, fuel consumption, and emissions. This knowledge is essential for engineers, mechanics, and enthusiasts aiming to optimize engine performance while adhering to environmental regulations. Join us as we unravel the complexities of engine performance and efficiency, illuminating the path to mastering the art of balancing power and environmental responsibility in internal combustion engines.

Chapter 9

Advancements and Future Trends

The field of internal combustion engines continues to evolve rapidly, driven by advancements in technology, environmental concerns, and the quest for higher efficiency. In this chapter, we will explore the latest innovations and future trends shaping the landscape of combustion engine technology.

9.1 Hybridization and Electrification

Combining internal combustion engines with electric motors has led to the development of hybrid vehicles. We will discuss hybridization techniques, regenerative braking, and the synergy between internal combustion and electric power. Additionally, we will explore fully electric vehicles, their challenges, and advancements in battery technology.

9.2 Downsizing and Turbocharging

To improve fuel efficiency without sacrificing power, engines are being downsized and equipped with turbochargers. We will delve into the principles of turbocharging, exploring advancements such as twin-scroll turbochargers and electric turbochargers. Downsizing strategies and their impact on engine performance will be analyzed.

9.3 Variable Valve Timing and Cylinder Deactivation

Variable valve timing (VVT) systems optimize engine performance under different operating conditions. Cylinder deactivation technology selectively shuts down cylinders to conserve fuel during light-load conditions. We will explore these technologies, their implementation, and their impact on overall engine efficiency.

9.4 Alternative Fuels and Sustainability

The search for sustainable fuels is driving research into biofuels, synthetic fuels, and hydrogen-based fuels. We will discuss these alternative fuel sources, exploring their production methods, environmental impact, and compatibility with internal combustion engines. Additionally, we will examine the concept of carbon-neutral combustion and its potential implications.

9.5 Advanced Materials and Manufacturing Techniques

Advancements in materials science and manufacturing processes have led to the development of lightweight and durable engine components. We will explore materials such as carbon fiber, ceramics, and advanced alloys, discussing their applications in engine design. Additionally, innovative manufacturing techniques like 3D printing are reshaping the production of engine parts.

By the end of this chapter, readers will gain insights into the cutting-edge technologies and future directions of internal combustion engines. Understanding these advancements is crucial for engineers, researchers, and policymakers involved in the automotive and transportation industries. Join us as we explore the forefront of combustion engine technology, shedding light on the innovations that will drive the future of sustainable and efficient transportation.

Chapter 10

Environmental Impact and Regulations

As concerns about environmental sustainability grow, internal combustion engines face increasing scrutiny due to their emissions and impact on the environment. In this final chapter, we will examine the environmental challenges posed by combustion engines and the regulations and technologies aimed at mitigating their impact.

10.1 Air Pollution and Greenhouse Gas Emissions

Combustion engines release pollutants such as nitrogen oxides (NOx), particulate matter (PM), and carbon dioxide (CO2) into the atmosphere. We will explore the environmental consequences of these emissions, including their contribution to smog, acid rain, and climate change. Understanding the sources and effects of these pollutants is essential for addressing environmental concerns.

10.2 Regulatory Standards and Emission Control Technologies

Governments and international organizations have implemented stringent emission standards to curb pollution from internal combustion engines. We will discuss regulatory bodies such as the Environmental Protection Agency (EPA) and the Euro standards, exploring the evolution of emission limits over time. Additionally, we will examine emission control technologies, including catalytic converters, diesel particulate filters (DPF), and selective catalytic reduction (SCR) systems.

10.3 Sustainable Practices and Alternative Fuels

To reduce the environmental impact of combustion engines, the adoption of sustainable practices and alternative fuels is crucial. We will explore biofuels, hydrogen, and synthetic fuels, discussing their potential as low-carbon or carbon-neutral alternatives. Additionally, we will examine sustainable mobility solutions such as electric vehicles, public transportation, and active transportation modes like cycling and walking.

10.4 Technological Innovations and Future Prospects

Technological innovations, such as predictive engine control algorithms and real-time emissions monitoring, hold the promise of further reducing emissions from internal combustion engines. We will explore these advancements and their potential to enhance engine efficiency and minimize environmental impact. Furthermore, we will discuss the prospects of emerging technologies, such as hydrogen fuel cell vehicles and advanced carbon capture methods.

By the end of this chapter, readers will have a comprehensive understanding of the environmental challenges associated with combustion engines and the strategies employed to address these issues. Awareness of these challenges and solutions is essential for policymakers, engineers, and consumers working collaboratively to create a sustainable future for transportation. Join us as we explore the environmental impact of combustion engines and the innovative solutions that aim to minimize their footprint on our planet.

Conclusion

The Evolution and Future of Combustion Engines

The internal combustion engine, a marvel of engineering ingenuity, has shaped the course of modern civilization, revolutionizing transportation, industry, and our way of life. From the early days of steam-powered machines to the sophisticated petrol and diesel engines of today, the evolution of combustion engines stands as a testament to human innovation and adaptability.

Throughout this comprehensive exploration, we have delved into the intricate workings of internal combustion engines, unraveling their fundamental principles, components, and processes. We have examined the engine cycles that drive their power, the meticulous engineering behind their components, and the sophisticated systems that regulate their temperature and lubrication. We have also scrutinized the combustion process, understanding the delicate balance between efficiency, power, and environmental impact.

In the pursuit of greater efficiency and reduced emissions, the landscape of combustion engines is undergoing a transformative shift. Hybridization, electrification, and alternative fuels are reshaping the industry, propelling us toward a future where internal combustion engines coexist with sustainable technologies. The integration of advanced materials, precise control systems, and predictive algorithms is ushering in an era of unprecedented performance and environmental responsibility.

However, challenges persist. Striking the delicate balance between power and environmental conservation remains a constant endeavor. As emissions regulations tighten and environmental awareness deepens, the need for innovation and collaboration becomes more pressing than ever. The journey toward cleaner, more efficient combustion engines demands the collective efforts of scientists, engineers, policymakers, and consumers alike.

In conclusion, the story of combustion engines is one of continuous evolution, adaptation, and resilience. As we stand on the cusp of a new era in transportation, the lessons learned from the past, coupled with the relentless pursuit of innovation, will guide us toward a future where internal combustion engines are not just powerful and efficient but also environmentally responsible. The legacy of the combustion engine is far from over, it is a legacy of innovation, sustainability, and the unwavering human spirit to overcome challenges and pave the way for a brighter, greener future.

www.ingramcontent.com/pod-product-compliance
Lightning Source LLC
Chambersburg PA
CBHW071005260726
48661CB00007B/2797